Education 143

不要恐慌

Don't Panic!

Gunter Pauli

冈特·鲍利 著

凯瑟琳娜·巴赫 绘

何家振 译

特别感谢以下热心人士对译稿润色工作的支持：

王必斗　王明远　王云斋　徐小怙　梅益凤　田荣义

乔　旭　张跃跃　王　征　厉　云　戴　虹　王　逊

李　璐　张兆旭　叶大伟　于　辉　李　雪　刘彦鑫

刘晋邑　乌　佳　潘　旭　白永喆　朱　廷　刘庭秀

朱　溪　魏辅文　唐亚飞　张海鹏　刘　在　张敬尧

邱俊松　程　超　孙鑫晶　朱　青　赵　锋　胡　玮

丁　蓓　张朝鑫　史　苗　陈来秀　冯　朴　何　明

郭昌奉　王　强　杨永玉　余　刚　姚志彬　兰　兵

廖　莹　张先斌

目录

Contents

蚂蚁一家看着那些忙忙碌碌的白蚁，被他们的纪律和秩序打动了。

“你们在这热带草原建造了多么壮观的装饰艺术天际线啊！” 一只蚂蚁说道。

“嗯，这是我们做需要做的事——保持生态系统健康。”白蚁答道。

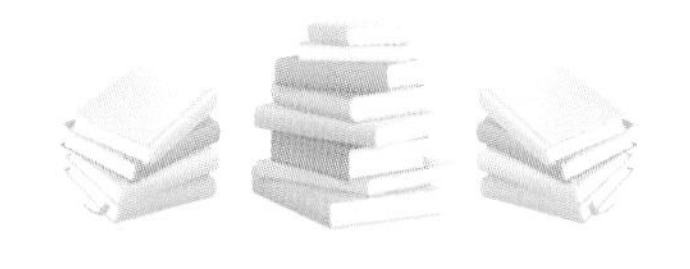

A family of ants is observing some termite friends at work, and are struck by their discipline and order.

“What an extraordinary Art Deco skyline you have built out here in the savannah!” one ant remarks.

“Well, we do what we need to do – to keep the ecosystem healthy,” Termite replies.

蚂蚁一家看着那些忙忙碌碌的白蚁……

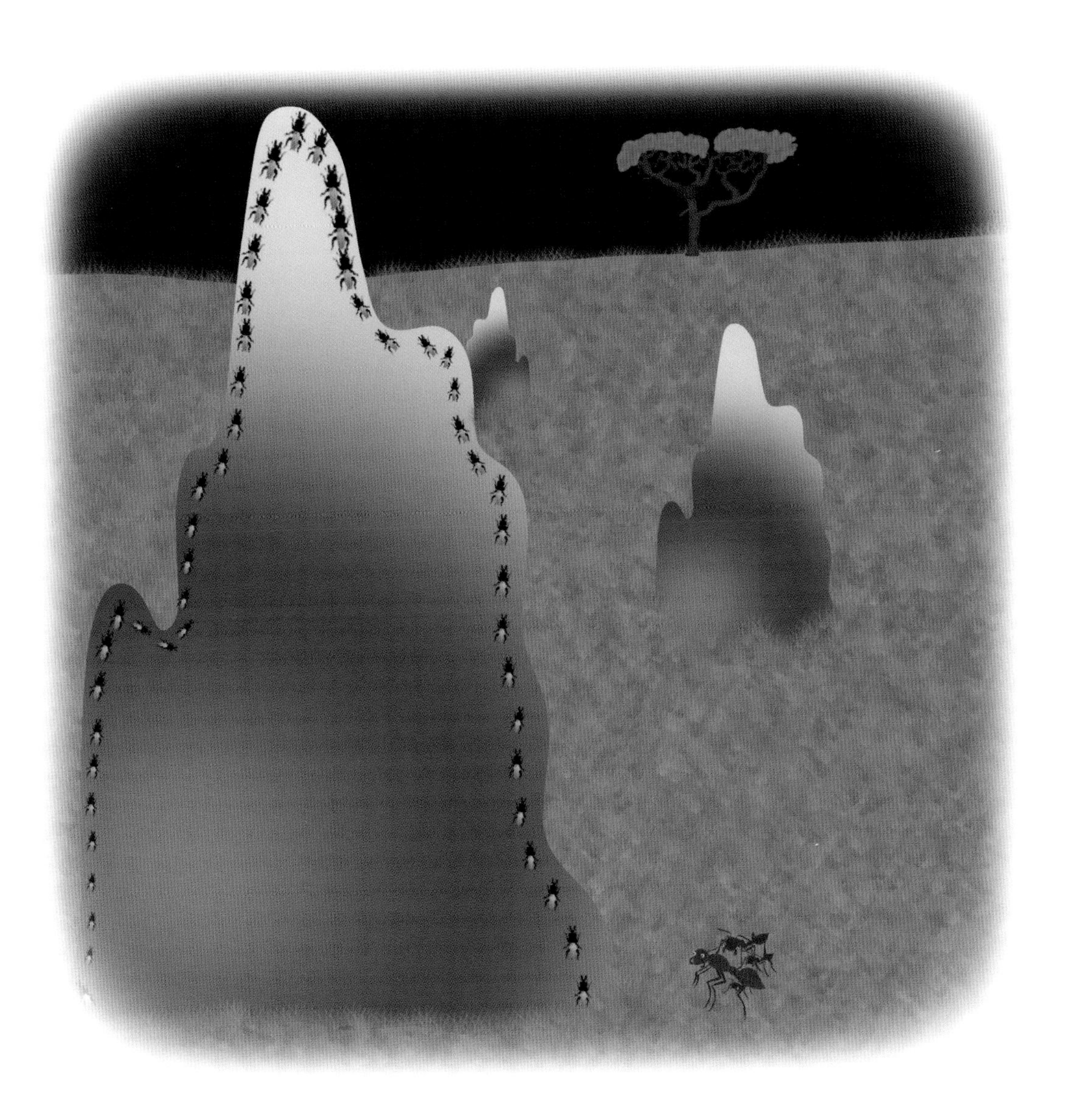

A family of ants is observing termites ...

在沙漠里创造出绿洲

Create an oasis in the desert

“我原本以为，你们所有的工作就是种植蘑菇。”

“哦，我们也种蘑菇。我们必须在所有地方都努力工作——在沙漠里、热带雨林里和大森林里。我们根据当地的需要调整我们的工作。但是那不是全部！我们甚至能在沙漠里创造出绿洲。”

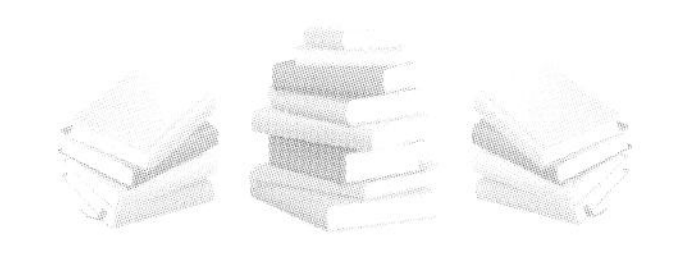

“And I thought mushrooms farming was all you did.”

“Oh, we do that, too. We do have to work really hard everywhere – in the desert, the rain forest and in woodlands. We adjust to the local need everywhere. But that is not all! We can even create an oasis in the desert.”

“我听说过你们的本事。你们真的能阻止土地干涸，变成荒原吗？”蚂蚁问道。

“瞧，我们所做的叫作播种。这使土壤变得健康、肥沃。通过保持土壤湿润，防止极端气候变化对土壤的破坏。”

“如果你们对环境那么好，那我就奇怪了，为什么人们认为你们是害虫——那种会吃掉他们房子上木头的害虫呢？”

“I heard that about you. Are you really able to prevent land from drying out and becoming waste land?” Ant asks.

“Look, what we are doing is called seeding. It makes the soil healthy and fertile. By keeping the soil moist, it is protected against drastic climate changes.”

“If you are that good for the environment, I wonder why people consider you a pest – one that can eat away at the timber of their houses?”

使土壤变得健康、肥沃！

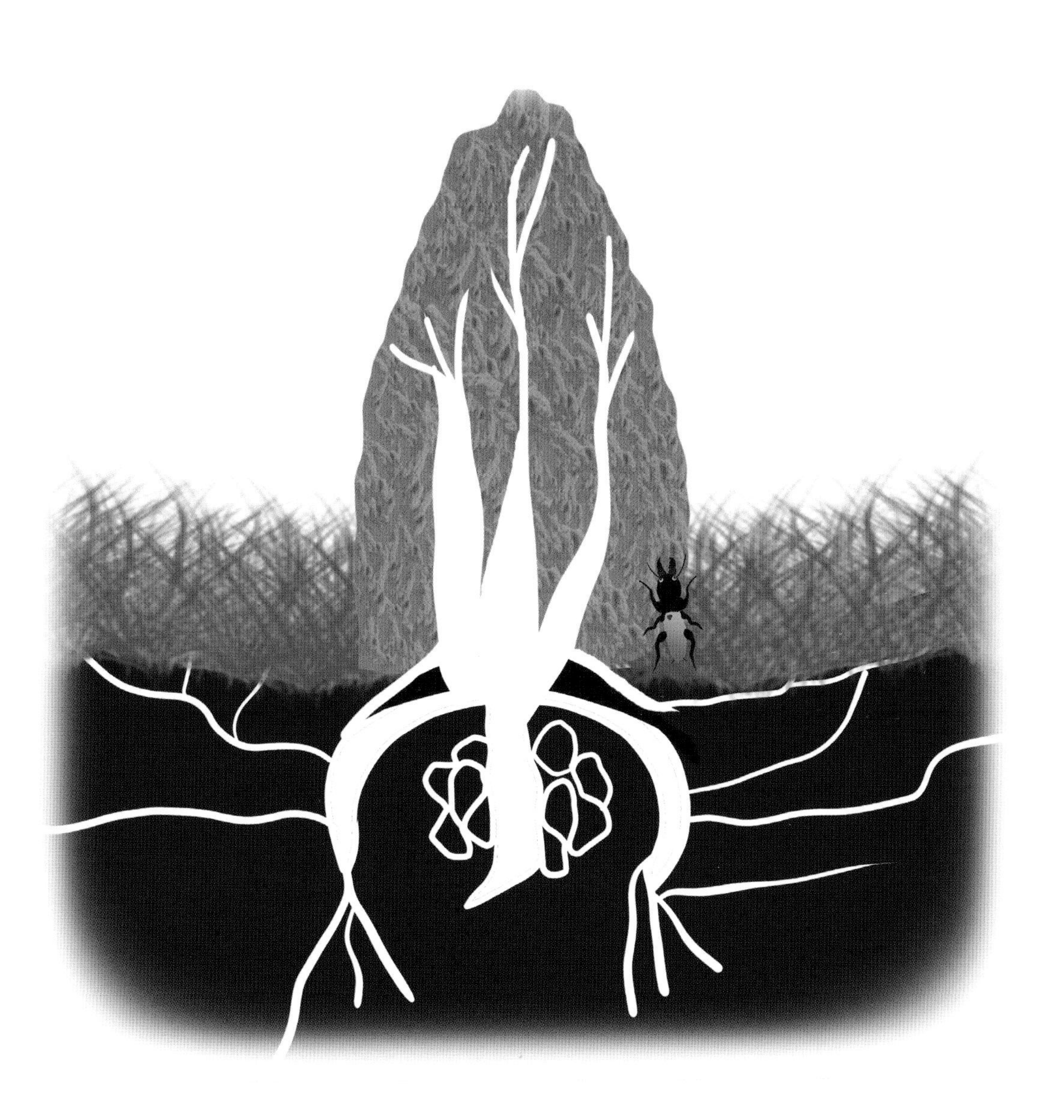

Makes the soil healthy and fertile!

人们能够更多地了解你们所做的有益工作

People can learn all the good you do

“听着，每个物种中都有害群之马，但这不意味着所有白蚁都是害虫！”

“但是人们知道这些吗？如果你们宣传得好一点也许对你们更有帮助。或许你们应该发起一次公关活动，这样人们能够更多地了解你们和你们所做的有益工作。”蚂蚁建议道。

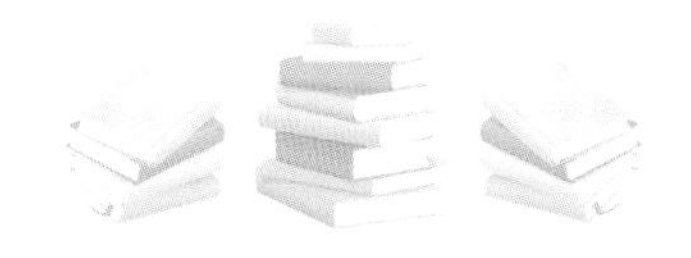

“Listen, there may be a rotten apple in every family, but that does not make all termites pests!”

“But do people know this? It may help if you communicated a little better. Perhaps you should launch a public relations campaign, so that people can learn more about you and the good work you do,” Ant suggests.

“我们不会在报纸上做广告，也不会在盛大的电视表演节目里打广告。我们只做我们做的事。如果人们愿意近距离观察我们，他们将能说服自己——而不用我们试图这样做！”白蚁回答道。

“把你们做的好事列个清单怎么样？你知道人类多么喜欢在互联网上搜索信息吗？这样他们就能更快地了解你们了。”蚂蚁提出另外一个建议。

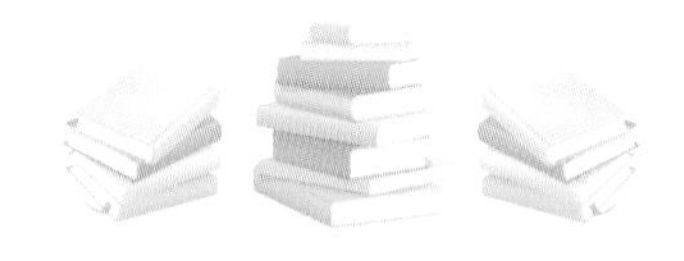

“We don’t place adverts in newspapers, or put on dog and pony shows on TV. We just do what we do. If people care to observe us closely, they will convince themselves – instead of us trying to do so!” Termite replies.

“What about making a list of all the good things you do? You know how people like to look up things on the internet? This way they can quickly learn more about you,” Ant puts forward another suggestion.

我们不会做广告……

We don't place adverts ...

好的土壤工程师……

Good soil engineers ...

“他们所要的只是一时之欢，那是不能持久的！如果你自己发现了什么东西，你就会记住它。如果你自己经历了某件事情，你就会理解它。那才是真正的知识！”

“通过对你的观察，我明白了你们是优秀的土壤工程师，在土壤里捅出小孔让空气进入，同时混入有机质和无机质颗粒，改善了土壤结构。你内脏里的细菌也是最棒的固氮器。”

“What they want is instant gratification, and you know that doesn’t last! Now if you discover something for yourself, you will remember it. And if you experience something for yourself you will understand it. That is true knowledge!”

“By observing you I see you are good soil engineers, poking holes in the soil to let air in, mixing organic and inorganic particles and providing structure to soil. The bacteria in your gut are also the greatest fixers of nitrogen.”

“我也听说了你做的好事。像我们和蜜蜂一样，你们是关注群体的社会性昆虫。人类似乎通常只关心他们自己。当人们惊慌失措、四处乱跑时，一些人甚至会被杀死！”

“嗯，我也容易恐慌。当我们的巢穴被攻击时，我们蚂蚁也会匆忙乱跑，连推带挤，甚至从落下的蚂蚁身上翻过。”

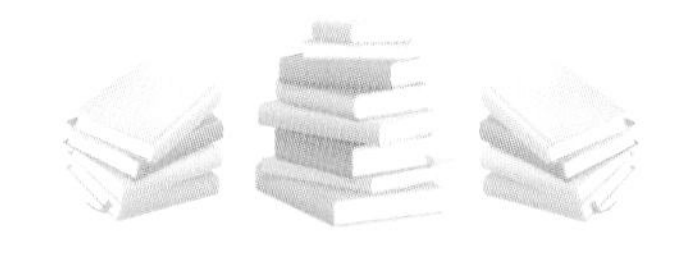

“And I have heard good things about you too. Like us, and bees, you are social insects, who care about their communities. People mostly seem to only care about themselves. When people panic and stampede some are even killed!”

“Well, I panic easily too. When our nest is attacked, we ants also run helter-skelter, push and shove, even clambering over the fallen.”

……甚至从落下的蚂蚁身上翻过！

… even clambering over the fallen!

文明的物种从来不践踏任何人……

Civilized beings never trample anyone...

“没有必要恐慌。”白蚁用平静而坚定的语气说。

“说得容易做着难！”

“如果被绊倒或者慢下来，其他的白蚁就会等他爬起来再走。文明的物种从来不践踏任何人……”

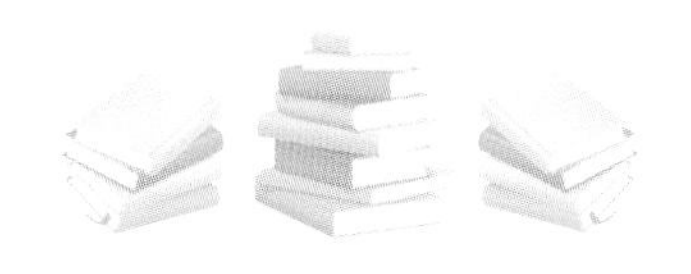

“There is no need to panic,” Termite says in a calm, determined tone.

“Easier said than done!”

“If a termite stumbles or slows down, the others simply wait for it to get going again. Civilized beings never trample anyone...”

“我不知道你们是如何做到的。恐怕我们蚂蚁恐慌的时候仍然是自私的。”

“那可能是因为我们白蚁开始家族群居生活比你们蚂蚁早数百万年吧。你们——就像人类一样——在彼此照顾方面还有很多东西要学。”

“噢，谢谢你们给我们树立了好榜样。”蚂蚁说。

……这仅仅是开始！……

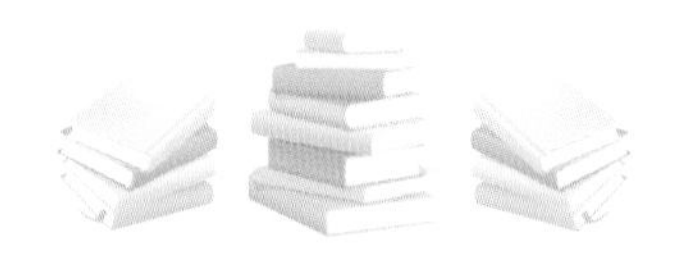

“I don’t know how you manage to do that. I’m afraid us ants are still very selfish when we panic.”

“It may perhaps be because we termites started living together as families millions of years before you ants did. You – just like people – still have a lot to learn about taking good care of each other.”

“Well, thank you for setting such a good example,” Ant says.

... AND IT HAS ONLY JUST BEGUN!...

……这仅仅是开始！……

... AND IT HAS ONLY JUST BEGUN! ...

Did You Know?

你知道吗？

Termites have lived in colonies for over 200 million years, about 100 million years longer than ants and bees have. When forced to flee a nest, the termite queen will carry mushroom mycelium with her to start a new food web.

白蚁群居生活2亿多年了，大约比蚂蚁和蜜蜂早1亿年。当被迫从巢穴逃离时，蚁后会带着蘑菇菌丝体以建立一个新的食物网。

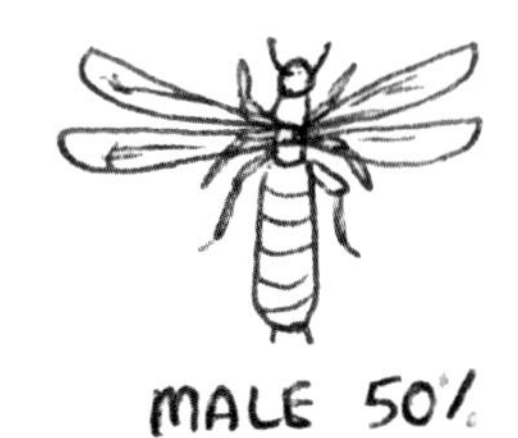

A termite colony has 50% males and 50% females, whereas a colony of ants or bees have mainly sterile females.

白蚁巢穴里有50%的雄性和50%的雌性，而蚂蚁和蜜蜂的巢穴里主要是不能生育的雌性。

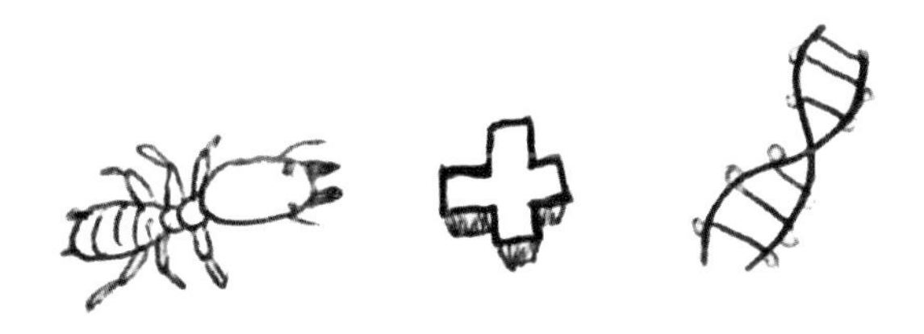

Termites focus on the genetic legacy of their colony, instead of only on their own offspring. Termites care for the vulnerable, especially in times of stress.

白蚁关注的是群体的基因遗传，而不是只关注它们自己的后代。白蚁照顾那些脆弱的个体，特别是在发生紧急情况时。

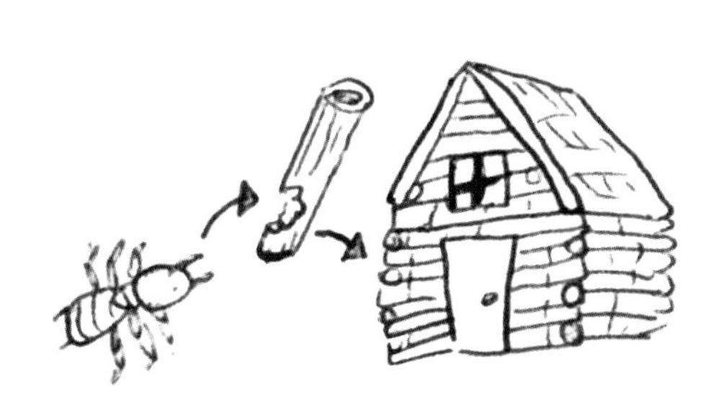

There are 3 000 families of termites and only a few feed on the timber of houses. As a biomass, termites weigh one hundred times what ants weigh.

白蚁大约有3000种，只有少数几种以吃房屋里的木材为生。从生物量来看，白蚁生物量是蚂蚁生物量的一百倍。

Termites have the greatest gut microbial density of all species. This enables them to eat what no other creatures can: wood, dung, lichen and dirt.

在所有物种中，白蚁肠道微生物密度最大。这让它们能吃其他生物不能吃的食物：木头、粪便、青苔和泥土。

Termites were the world's first 'farmers', and the most sustainable ones. Partnered with fungi, and providing them with a clean, safe, temperate and well-ventilated environment, free of competition, with enough water and food, the termites receive abundant food in return.

白蚁是世界第一“农夫”，而且是最可持续发展的农夫。作为蘑菇的伙伴，白蚁给蘑菇提供了清洁、安全、温和而又通风良好的环境，使其免于与其他物种竞争，有足够的水和食物，作为回报，白蚁也得到了丰富的食物。

The largest termite mounds can be up to 10 m high and 25 m wide, housing 10 million termites. These termite hills secure the long-term fertility of the land, putting up to 15% of available biomass back into the soil.

最大的白蚁丘可高达 10 米，宽 25 米，容纳 1000 万只白蚁。白蚁丘的存在保证了土地的长期肥力，给土壤回补了 15% 的生物量。

The inside of a termite mound is cool during the heat of the day, and warm on a cold night. For termites, it is a pleasant place to be in.

在炎热天气里，白蚁丘内部凉爽，但是到了寒冷的夜晚，白蚁丘又很温暖。对白蚁来说，这是一个舒适的居住地。

Think About It 想一想

What would you do if there were a panic in a room or a stadium: run whatever the cost or risk?

如果在房间或体育场内发生了让人恐慌的事件，你会不惜代价和风险逃跑吗？

Do you like the idea of termites 'farming' by creating a system where they grow mushrooms in their mounds and get to eat some of the prolifically growing mushrooms?

白蚁在它们的蚁丘里种植蘑菇，并吃掉一部分大量生长的蘑菇。你喜欢白蚁通过创建一个系统而成为农夫的主意吗？

Are we too quick in judging a whole family or an entire species because one of their members does things we do not like?

仅仅因为其中的一个物种或门类中的一个成员做了我们不喜欢的事情，就给整个物种或门类下结论，是不是太草率了呢？

How fast do you want answers to your questions? Are you prepared to take the time to find things out for yourself?

你会迫不急待知道问题的答案吗？还是更愿意自己花费一些时间去探索事情的原委呢？

What can you do to improve soil fertility? Think about the fact that all our food, everything we eat, somehow needs nutrition extracted from the soil that forms the top layer of the Earth. We need to ensure that this remains available for centuries, even millennia, to come. While it is easy to work on the surface adding fertilisers, how can we go about contributing to the fertility of the deeper layers? Do some research and draw up a list of possible initiatives. Now test these as concepts.

为了改善土壤肥力你能做什么？想一想，我们所有的食物、我们吃的所有东西，都需要从地球表面最上层的土壤中汲取营养物质。我们需要确保这种耕作方式能够在未来的几个世纪甚至几千年里持续。虽然在表层施肥做起来不难，但是深层土壤施肥该怎样做呢？做一些研究并列出可能的办法。测试一下这些办法。

学科知识

Academic Knowledge

生物学	肠道里细菌的多样性；微生物帮助消化。
化　学	纤维素通过真菌分解，为深层土壤提供肥力；纤维素发酵产生热量；土壤中的空气流通对保持有氧环境的重要性；可以利用白蚁固氮。
物　理	白蚁有控制巢穴里湿度和温度的能力；利用烟囱原理排出蚁穴中的空气，并在穴内造成一个低压系统，引导外部空气通过通道进入蚁穴。
工程学	相对它们自身体积，白蚁建造了世界上最大的工程建筑；白蚁利用太阳能维持蚁穴内温度稳定。
经济学	白蚁为土壤提供长期肥力，持续时间超过一个季度甚至一年，从长远的角度看，白蚁的存在非常有必要；有些白蚁会对建筑物造成危害。
伦理学	如果只是个别害群之马做的坏事，我们怎么能够把整个种群都看成有害的呢？
历　史	白蚁是地球上最古老的群居生物，已经存在了2亿多年；白蚁可能是地球上最早的农夫；装饰艺术时代起始于1925年，在20世纪30年代成为主流。
地　理	最高的白蚁丘在非洲，高达10米。
数　学	穴内与穴外的温度，低压，空气通道的长度以及湿度之间的关系，确保了蘑菇的持续生产，为多达千万白蚁种群提供食物，这表明白蚁精通数学。
生活方式	多数人只关注短期生存，不考虑子孙后代的发展；我们已经习惯于即时满足：利用即时的信息来源，寻找或购买我们想要的东西；更多地关心自己，而很少关心他人；我们需要不断学习新知识。
社会学	尽管只是少数成员造成的破坏，却使整个部落或者家族遭到谴责；发生恐慌时的惊跑可能会造成踩踏事件；社会如何采取预防措施保证不发生损害；大合唱的重要性；在非洲和澳大利亚白蚁丘过去常常被称为蚂蚁山，尽管没有蚂蚁住在里边，蚂蚁山现在被称为白蚁穴。
心理学	我们简单地相信广告里的所见所闻，把电视节目与现实混淆；相信那些似是而非的推送消息；当紧急和恐慌事件发生时，如何能保持冷静并且帮助弱者；如何用纪律约束自己；从自身的经历中，提炼形成一个能牢记的知识点；你自己发现的东西会记得更牢。
系统论	白蚁通过向表层土回补15%的生物量，保证了土壤的长期肥力。

情感智慧

Emotional Intelligence

蚂 蚁

蚂蚁欣赏热带草原地平线上由众多白蚁丘形成的视觉效果，并表现出对白蚁的同理心。她承认她对白蚁在自然界里角色的了解是很粗略的，并且质疑白蚁作为紧张的生态区域调节者的功能，她新得到的信息使她感到困惑。她不理解为什么人类把白蚁视为问题。她很自信地向白蚁提出了一个多层面的宣传战略，以让人们更好地了解他们在自然中的作用。尽管白蚁拒绝了她的一个又一个建议，她仍不断地修改自己的行动计划。蚂蚁认识到白蚁完成了不同角色的使命是如此伟大。这不仅表现在他们的工作中，也表现在他们遇到紧急情况的时候。

白 蚁

白蚁很谦逊，不认为他所做的是一项非凡的工作。他指出在世界各地还有很多工作需要做。白蚁不仅仅是完成一项任务；他以防止破坏生态系统为目标，胸怀长远梦想，做好自己的工作。他承认一些白蚁家族的成员的确造成破坏，但是拒绝那种以少数成员的行为判断整个物种的观点。白蚁没有寻求引人注目的焦点，而是更倾向于自己作为一个物种被发现。白蚁没有接受赞扬，而是解释了白蚁如何保持冷静，如何关照他们中的弱者，给所有的白蚁都带来安全。白蚁还分享了他的终生学习的生活方式。

艺术

The Arts

社区是有很强社会凝聚力、在遭遇危机时互相帮助的一群人。确保社区里的每个人在紧急情况下都知道该如何做的方法之一，是定期对各种紧急情况进行模拟演练，以便训练社区成员在遇到紧急情况下保持冷静，并采取明智的行动。建立社区的另一种方式是大家一起唱所有人都会唱的流行歌曲。它可以是国歌、校歌或者摇滚金曲。和几个朋友一起，选择一首让你们团结起来的歌，然后演唱它。

思维拓展

Systems: Making the Connections

白蚁总是被认为是有害的昆虫。在3000多类白蚁种群中，只有3种已知对建筑物木材有害。很少有人认识到，白蚁通过给土壤补充肥力对自然循环做出的卓越贡献。据估计，每年有高达15%的生物量被白蚁耕耘返回土壤。白蚁内脏中种类繁多的细菌群落使它几乎能够消化所有的东西。这并非白蚁的唯一贡献。它们还会引起生物量的发酵和发热，可以帮助维持蘑菇群落繁荣（蘑菇是白蚁的主要营养食物来源）。同时也给植物的根部提供持续的水分和热量，刺激植物生长。这些为生态系统服务的功能，人们不愿意承认，大多数农民置白蚁的长期功能于不顾，指责白蚁对庄稼的毁坏和对叶子的吞食。人们已经着手完全消灭白蚁的行动，但是达成这个目标的概率很小。如果一个人在某个地区成功消灭了白蚁，那么那个地区的土壤很快就会变得贫瘠，这是不能依靠简单施加堆肥或使用化肥弥补的。对白蚁功能了解的缺乏是人们试图一劳永逸地摧毁和根除白蚁的所有努力都失败了的原因。一旦一个白蚁丘被毁，那么用不了多久白蚁就会建立一个新的聚居地。当灾难来袭时，蚁后将会尝试带着在她口中安全地储藏着的真菌孢子逃生。利用这些孢子，她会在另外一个巢穴建设一个新的蘑菇农场，在这里，幸存者们将会重建地下生产单元——包括帮助空气流通的通道，为昆虫和蘑菇群落的繁荣提供氧气和氮。

动手能力

Capacity to Implement

如何能在众多人恐慌时保持冷静呢？我们生活的世界确实会发生一些问题——从自然灾害，比如洪水、地震或海啸，到人为灾难，如在大型体育赛事里的惊跑，或者火灾。在这类事件中，如何把我们的家人、朋友以及其他人带到安全地带。为了避免在此种情况下只想到我们自己，一个最常用的办法是定期举行演练。你们学校有消防演习吗？你知道如何在危机发生期间照管弱者了吗？你和你的朋友在危机发生时学会照顾好自己和他人是非常重要的。

故事灵感来自

丹尼拉·鲁斯
Daniela Rus

丹尼拉·鲁斯拥有康奈尔大学计算机科学专业博士学位。在去麻省理工学院（MIT）担任电子工程和计算机科学教授之前，她是达特茅斯学院计算机科学系的教授。她也是计算机科学和人工智能实验室（CSAIL）主任。她的研究小组开发出自重构机器人、自组织机器人、移动传感网络和水下合作机器人技术。她的团队建造的机器人可以护理花园，烤饼干，切生日蛋糕，在没有人类帮助的情况下成群飞翔执行监控任务，而且可以与人跳舞。她的灵感来自不会发生恐慌的白蚁。

图书在版编目（CIP）数据

不要恐慌：汉英对照 /（比）冈特·鲍利著；（哥伦）凯瑟琳娜·巴赫绘；何家振译．— 上海：学林出版社，2017.10
（冈特生态童书．第四辑）
ISBN 978-7-5486-1267-4

Ⅰ．①不…　Ⅱ．①冈…　②凯…　③何…　Ⅲ．①生态环境－环境保护－儿童读物－汉、英　Ⅳ．① X171.1-49

中国版本图书馆 CIP 数据核字（2017）第 143509 号

著作权合同登记号　图字 09-2017-532 号

冈特生态童书
不要恐慌

作　　者—— 冈特·鲍利
译　　者—— 何家振
策　　划—— 匡志强 张　蓉
责任编辑—— 许苏宜
装帧设计—— 魏　来
出　　版—— 上海世纪出版股份有限公司学林出版社
地 址：上海钦州南路 81 号　电话 / 传真：021-64515005
网 址：www.xuelinpress.com
发　　行—— 上海世纪出版股份有限公司发行中心
（上海福建中路 193 号　网址：www.ewen.co）
印　　刷—— 上海丽佳制版印刷有限公司
开　　本—— 710×1020　1/16
印　　张—— 2
字　　数—— 5 万
版　　次—— 2017 年 10 月第 1 版
2017 年 10 月第 1 次印刷
书　　号—— ISBN 978-7-5486-1267-4/G.493
定　　价—— 10.00 元

（如发生印刷、装订质量问题，读者可向工厂调换）